From Your Friends at *The Mailbox*

Geography
MIND BUILDERS
Grades 4–6

Welcome to *Geography Mind Builders*! This must-have resource is sure to reinforce geography and map skills while developing critical-thinking skills. Packed with challenging questions covering United States and world geography, this resource provides students with a school year's worth of mind-building opportunities.

Project Manager:
Irving P. Crump

Writer:
William Fitzhugh

Art Coordinator:
Pam Crane

Artists:
Pam Crane, Teresa R. Davidson

Cover Artists:
Nick Greenwood, Clevell Harris, Kimberly Richard

www.themailbox.com

©2001 by THE EDUCATION CENTER, INC.
All rights reserved.
ISBN #1-56234-439-0

Except as provided for herein, no part of this publication may be reproduced or transmitted in any form or by any means, electronic or mechanical, including photocopying, recording, or storing in any information storage and retrieval system or electronic online bulletin board, without prior written permission from The Education Center, Inc. Permission is given to the original purchaser to reproduce patterns and reproducibles for individual classroom use only and not for resale or distribution. Reproduction for an entire school or school system is prohibited. Please direct written inquiries to The Education Center, Inc., P.O. Box 9753, Greensboro, NC 27429-0753. The Education Center®, *The Mailbox*®, and the mailbox/post/grass logo are registered trademarks of The Education Center, Inc. All other brand or product names are trademarks or registered trademarks of their respective companies.

Manufactured in the United States
10 9 8 7 6 5 4 3 2 1

INCLUDED IN THIS BOOK

Map Skills 3–15

United States Geography 16–29

World Geography 30–42

Answer Keys 43–48

HOW TO USE THIS BOOK

Inside this resource you will find a variety of questions designed to reinforce the geography topics and skills that you teach. Each activity page features five mind-building geography questions, plus a more difficult bonus builder question to boost students' critical-thinking and research skills—a total of 240 questions in all!

Use the pages in this book in a variety of ways to supplement your social studies curriculum:

For independent practice, duplicate the pages for students to use as morning work, free-time activities, or homework.

Use the pages as resources for questions of the day. Make and display a transparency of each page, revealing one question each day for students to answer. At the end of the week, answer and discuss all five questions. Assign that page's bonus builder as weekend homework.

For partner or small-group research, duplicate the desired page and give each pair or group a copy.

For a whole-group activity, select the cards that relate to your current topic of study. For example, when studying North America, pull all of the cards that relate to Canada, Mexico, Central America, Greenland, the Caribbean Islands, and the United States.

For a learning center activity, duplicate, laminate, and cut apart the pages. Group the resulting cards by topic and place specific cards at a center.

Note: Information contained in this book is current as of November 2000. Questions dealing with population statistics are based on the 1990 census or on census projections for 2000.

Map Skills

What island country lies off the southeast coast of India? What was this country once called?

(1)

Map Skills

What large central European country was two separate countries from 1949–1990? (These two countries were reunited in 1990.)

(2)

Map Skills

Which U.S. state extends farther west: Virginia or West Virginia?

(3)

Map Skills

Lines on a globe or map help us locate places. Name the imaginary line that
a. circles the middle of the earth
b. separates one day from the next
c. marks 0° longitude
d. separates the Tropics from the temperate zone in the Southern Hemisphere

(4)

Map Skills

Much of western Australia is made up of _____.

(5)

Bonus Builder #1

Which of the following world cities are located in the Tropics?
a. Darwin, Australia
b. Miami, Florida
c. Buenos Aires, Argentina
d. Honolulu, Hawaii
e. Cairo, Egypt
f. Beijing, China

Map Skills

©2001 The Education Center, Inc. • Mind Builders • Geography • TEC1609 • Key p. 43

Map Skills

Earth is a planet of extremes.
a. The highest place on Earth is _____.
b. The lowest place on Earth's land surface is _____.
c. Both of these places are located on the continent of _____.

(6)

Map Skills

A *tributary* is a small river that flows into a larger one. List five tributaries of the Mississippi River.

(7)

Map Skills

Which city is farther west: Carson City, Nevada, or San Diego, California? How do you know?

(8)

Map Skills

Canals are important in helping move goods from one place to another. Name the location of each of the following canals:
a. Suez Canal
b. Panama Canal
c. Erie Canal

(9)

Map Skills

The Tennessee Valley Authority brought electricity, flood control, and better river transportation to many states. When was the TVA created? List five states that the TVA helped.

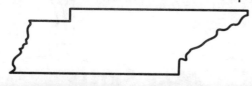

(10)

Bonus Builder #2

Which of the following African cities are north of the equator? Which ones are south?
a. Casablanca   d. Lagos
b. Dar es Salaam e. Algiers
c. Dakar   f. Johannesburg

Map Skills

Map Skills

Which of the following states border an ocean?
a. Wyoming
b. South Carolina
c. Alaska
d. Iowa
e. Florida
f. Maine
g. Minnesota
h. Idaho
i. California

(11)

Map Skills

The imaginary line halfway between the North Pole and the South Pole is the _____. The imaginary line located at 66.5° north latitude is the _____.

a. Which line is Hawaii nearer?
b. Which line is Alaska nearer?

(12)

Map Skills

Continents are large land masses.
a. Which continent has the largest land area?
b. Which continent has the largest population?
c. Which continent has no permanent population?

(13)

Map Skills

Which continents are located entirely in the Southern Hemisphere?

(14)

Map Skills

A map and a globe are both important to geographers. List two ways maps and globes are similar. List two ways they are different.

(15)

Bonus Builder #3

The *mouth* of a river is the place where a river flows into a larger body of water. Into which body of water does each of the following rivers flow?
a. Nile River
b. Murray River
c. Amazon River
d. Ganges River
e. Seine River

Map Skills

Map Skills

The United States and Canada share the longest undefended border in the world. List five Canadian provinces that border the United States.

16

Map Skills

Use a map to help you identify these European countries. Then list them in order from largest to smallest.

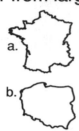
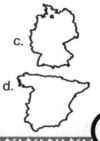

17

Map Skills

In which direction would you travel if you flew from Chicago to each of the following cities?
a. San Francisco
b. Philadelphia
c. New Orleans
What map element did you use to help you?

18

Map Skills

List the five U.S. states that border the Gulf of Mexico.

19

Map Skills

A waterfall is a magnificent sight! Where is each of the following waterfalls located?
a. Niagara Falls
b. Angel Falls
c. Victoria Falls

20

Bonus Builder #4

Some countries are islands. List an island country that is part of each of the following continents:
a. Europe
b. Asia
c. Africa

Map Skills

MAP SKILLS

A map's *scale* tells you the relationship between the real distance on the earth and the distance on the map. Below are some measurements made on a map. Use the scale to find the real distances on the earth.

a. 2 inches = _____ miles
b. 5 inches = _____ miles
c. 1/2 inch = _____ miles
d. 3 1/2 inches = _____ miles

Scale
1 inch = 40 miles

MAP SKILLS

What special symbol is often used to show a national capital on a map? Which of the following European cities are *not* national capitals?

a. Barcelona   d. Rotterdam
b. Lisbon   e. Brussels
c. Marseille   f. Oslo

MAP SKILLS

Scandinavia refers to three European countries. What are they?

MAP SKILLS

Which of the following countries are not located along the equator?

a. Brazil
b. Paraguay
c. Angola
d. Kenya
e. Indonesia
f. Japan

MAP SKILLS

What kind of area is warm all year long, usually with long periods of heavy rainfall? (Hint: The Amazon River Basin includes an example.)

BONUS BUILDER #5

Symbols help us use maps. What map feature does each symbol below usually stand for?

a.   c.

b.   d.

MAP SKILLS

Map Skills

Which continents are located entirely in the Northern Hemisphere?

26

Map Skills

Which African countries border the Mediterranean Sea?

27

Map Skills

I am the largest continent. I have the greatest population, too. What am I?

28

Map Skills

Name the location of each of the following straits:
a. Strait of Magellan
b. Strait of Gibraltar
c. Bering Strait
d. Strait of Dover

29

Map Skills

I am the world's coldest desert. I am also the northernmost desert in the world. I stretch for more than a half million square miles across part of southern Mongolia and northern China. What am I?

30

Bonus Builder #6

Physical maps, political maps, thematic maps—each type of map has a special use. Which kind of map would you use to learn about the following:
a. elevations   d. population
b. names of countries e. growing seasons
c. annual rainfall

Map Skills

Map Skills

What major mountain range extends from western Canada to the southwestern United States? List five states found in this mountainous region.

31

Map Skills

The Continental Divide is important in geography. Rivers to the west of the divide generally flow toward the Pacific Ocean; rivers to the east of the divide flow toward the east. List three rivers on the western side of the Continental Divide.

32

Map Skills

List three ways *rural* communities and *urban* communities are different from each other.

33

Map Skills

Name two ways *polar* regions and *deserts* are alike.

34

Map Skills

Water is usually blue on maps. Which of the following are examples of water?

    a. isthmus   d. archipelago
    b. strait   e. bight
    c. shield   f. steppe

35

Bonus Builder #7

The equator divides the earth into northern and southern hemispheres. In which hemisphere is each of the following cities?

    a. Buenos Aires   d. Sydney
    b. New York City   e. Madrid
    c. Manila   f. Shanghai

Map Skills

Map Skills

Cities often develop along important rivers. Name the river near each of the following cities:

a. London, England
b. New York, New York
c. Paris, France
d. New Orleans, Louisiana
e. Cairo, Egypt

36

Map Skills

Which of the following cities is closest to the Arctic Circle?

a. Toronto
b. Beijing
c. Reykjavik
d. Moscow
e. Vladivostok

37

Map Skills

Which body of water is larger: the Red Sea or the Mediterranean Sea? List the three continents that the Mediterranean Sea borders.

38

Map Skills

I am on the western coast of a large continent. The Andes Mountains form my eastern boundary. I am the world's leading copper producer. Most of my people speak Spanish. What am I?

39

Map Skills

Which country extends across the most longitudes: the United States, China, Brazil, or Russia?

40

Bonus Builder #8

Suppose you wanted to travel from London, England, to Bombay, India, by ship—and by the shortest possible route. Over which bodies of water would you sail?

Map Skills

Map Skills

The Falkland Islands, off the eastern coast of South America, belong to which European country?

41

Map Skills

The ocean current that helps to warm parts of western Europe is called the _____.

42

Map Skills

Many geographers consider the _____ to be one of the boundaries that separate Europe and Asia.

43

Map Skills

The Great Barrier Reef is found in the Coral Sea off the coast of _____.

44

Map Skills

If you flew from Rio de Janeiro, Brazil, to Santiago, Chile, over which mountains would you fly?

45

Bonus Builder #9

The country once known as Yugoslavia is now broken up into five smaller, independent nations. One of those five nations kept the name Yugoslavia and consists of Serbia and Montenegro. Name the other four nations.

Map Skills

MAP SKILLS

Landforms are natural features of a land surface. Which of the following are landforms?

a. tundra
b. mesa
c. strait
d. plateau
e. bight
f. island
g. gulf

(46)

MAP SKILLS

Which of the following islands are *not* found in the Caribbean Sea?

a. Cuba
b. Timor
c. Jamaica
d. Grenada
e. Dominica
f. Madagascar
g. Java

(47)

MAP SKILLS

Which African country has seacoasts on both the Atlantic and Indian oceans?

(48)

MAP SKILLS

My region covers parts of Morocco, Algeria, Tunisia, Libya, Sudan, Chad, Niger, Mali, and Mauritania. The average rainfall here is less than four inches a year. What am I?

(49)

MAP SKILLS

Is Africa greater in distance from east to west or from north to south? How can you prove your answer?

(50)

BONUS BUILDER #10

Bodies of water often separate land regions. Name the body (or bodies) of water that separates each of the following:

a. Madagascar—Africa
b. Alaska—Russia
c. Tasmania—Australia
d. England—France

MAP SKILLS

MAP SKILLS

When locating information on a map, it's important to use the map's elements. List six elements of a map.

51

MAP SKILLS

Which country in western Asia near the Black Sea has the same name as a U.S. state?

52

MAP SKILLS

Which North American country extends farthest east?

53

MAP SKILLS

To which European country does Greenland belong? Greenland is a part of which continent?

54

MAP SKILLS

What is the name of the waterway system that improved transportation between the Great Lakes and the Atlantic Ocean?

55

BONUS BUILDER #11

List at least five countries that are found in Europe's Balkan Peninsula.

MAP SKILLS

MAP SKILLS

Where do more than 80 percent of Australia's people live?

56

MAP SKILLS

Which southeast Asian country is divided into a "North" and a "South"?

57

MAP SKILLS

Large dots usually show the locations of large cities on a map. What are the three largest cities in California?

58

MAP SKILLS

What area of Australia receives the greatest annual rainfall?

59

MAP SKILLS

Which Canadian province is made up of part of the Canadian mainland and an island?

60

BONUS BUILDER #12

Name the largest country on each of the following continents:

    a. Asia   d. Africa
    b. South America   e. North America
    c. Europe

Why aren't Australia and Antarctica included?

MAP SKILLS

MAP SKILLS

The seasons north and south of the equator are opposite each other. What season is it now where you live? What season is it in each of the following cities?

   a. Montevideo, Uruguay
   b. Jerusalem, Israel
   c. Tokyo, Japan
   d. Buenos Aires, Argentina
   e. Sydney, Australia

61

MAP SKILLS

Name two ways deserts and mountains are alike. Name two ways they are different.

62

MAP SKILLS

Which city do you think has a warmer climate: Cape Town, South Africa, or Lagos, Nigeria? Why?

63

MAP SKILLS

I am an island nation located near the Arctic Circle. I am often called the "Land of Ice and Fire" because of my glaciers, steaming hot springs, geysers, and volcanoes. Although my name is "cold," I'm not as cold as most places so far north. What am I?

64

MAP SKILLS

Which continent has the most countries? Name the three largest countries on this continent.

65

Bonus Builder #13

The breakup of the former Soviet Union created many new country names on the map. Name five of these new countries.

MAP SKILLS

United States Geography

Before the beginning of the U.S. Civil War, which states were slaveholding states?

66

United States Geography

My state flower is the bluebonnet. Oklahoma is north of me. Mexico and the Gulf of Mexico are south of me. I am the second largest state. What am I?

67

United States Geography

Which two states are shaped like rectangles?

68

United States Geography

I helped Pierre L'Enfant and Andrew Ellicott draw up the plans for the new city of Washington, DC, over 200 years ago. Who am I?

69

United States Geography

In which states would you find the following man-made landmarks?

 a. Gateway Arch
 b. Faneuil Hall
 c. World Trade Center
 d. Independence Hall
 e. U.S. Naval Academy

70

Bonus Builder #14

Many colonies were begun by European settlers. Settlers from which country first settled each of the following states?

 a. New York   c. Pennsylvania
 b. Delaware   d. Florida

United States Geography

United States Geography

Many foods that Americans enjoy originally came from other countries. Name the country of origin for each of the following foods:

      a. pizza
      b. crepes
      c. tortillas
      d. sukiyaki

United States Geography

Many ships are important in American history. Write a sentence about each of the following ships. Include the place associated with the ship.

a. *Half Moon*
b. *Exxon Valdez*
c. *Lusitania*

United States Geography

I am home to Mount Rushmore, the Black Hills, and the Badlands. My state bird is the ring-necked pheasant. Minnesota is to the east of me. What am I?

United States Geography

Name the states whose outlines are shown below:

United States Geography

I am the largest lake in the Western United States. No fish live in my waters. My size varies according to weather conditions. What am I?

Bonus Builder #15

Five of the ten most populated U.S. cities are located in two states. Name those two states.

United States Geography

United States Geography

Write the state name that matches each two-letter postal abbreviation.

a. TN   d. CA
b. SC   e. AZ
c. MD   f. NY

What is *your* state's postal abbreviation?

United States Geography

The Appalachian Mountains extend from Canada through the eastern United States. Which of the following mountain chains is *not* part of the Appalachians?
a. Green Mountains
b. White Mountains
c. Blue Ridge Mountains
d. Ozark Mountains
e. Catskill Mountains
f. Allegheny Mountains

United States Geography

Which two states do not border any other states?

United States Geography

Unscramble each of the following state capitals:

a. NTUASI   d. YALABN
b. NTOBSO   e. EASLM
c. SOBIE   f. TATALNA

Now match each capital to its state.

New York   Massachusetts
Idaho   Georgia
Oregon   Texas

United States Geography

Many U.S. state capitals are named for famous people. List five state capitals named for people.

Which U.S. states were created from the Northwest Territory? Which of these states was the first to join the Union?

United States Geography

United States Geography

Three of the world's five largest natural lakes are found in North America. List them.

81

United States Geography

List four states that also have rivers with the same names.

82

United States Geography

We explored the vast Louisiana Territory for President Thomas Jefferson. We reached the Pacific Ocean in 1805. Who are we?

83

United States Geography

The Holocaust Memorial Museum is dedicated to remembering the millions of Jews and other minorities who perished at the hands of the Nazis in World War II. Where is this museum located?

84

United States Geography

The early American colonies participated in the *triangular trade.* Why was it called that?

85

Bonus Builder #17

If you look carefully at a group of five states in the middle of the United States, you'll see a shape that looks like a chef. This chef is nicknamed *MIMAL,* a word that is made up of the first letter of each of the five states. Which five states make up MIMAL?

United States Geography

UNITED STATES GEOGRAPHY

List the five most populous U.S. states.

(86)

UNITED STATES GEOGRAPHY

Which U.S. state is first in alphabetical order? Which state is last? What is the approximate distance between these two states' capital cities?

(87)

UNITED STATES GEOGRAPHY

List the following in order from least to greatest according to population:

a. North America   c. Memphis
b. Tennessee   d. United States

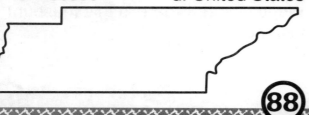

(88)

UNITED STATES GEOGRAPHY

We live in the Four Corners area of the southwest United States. Our tribe lives on a reservation. We are the second largest Native American tribe in the United States. Who are we?

(89)

UNITED STATES GEOGRAPHY

Which of the following cities are *not* located on the coast of an ocean?

a. Santa Fe   e. Boise
b. Los Angeles   f. Atlanta
c. Baltimore   g. New York City
d. New Orleans

(90)

Bonus Builder #18

Many books for young people are set in the United States. Which state is the setting for each of the following books?

a. *Johnny Tremain*
b. *Misty of Chincoteague*
c. *Caddie Woodlawn*

UNITED STATES GEOGRAPHY

United States Geography

Washington, DC, has many famous landmarks. Name the site where
a. the president lives
b. you would see the Tomb of the Unknowns
c. you would climb the city's tallest structure
d. Congress meets and makes laws

91

United States Geography

If you draw lines that connect Anchorage, Alaska, to Boston, Massachusetts, to Miami, Florida, and back to Anchorage, what shape will result? Between which two cities is there the greatest distance?

92

United States Geography

Which city is
a. farther west: Las Vegas, Nevada, or San Diego, California?
b. farther north: Portland, Maine, or Portland, Oregon?
c. farther south: San Diego, California, or Houston, Texas?

93

United States Geography

Match each state to its state bird:

a. chickadee   Arizona
b. nene   Idaho
c. brown pelican   Maine
d. cactus wren   Louisiana
e. mountain bluebird Hawaii

94

United States Geography

Harriet Tubman led runaway slaves north through the eastern United States to freedom in Canada. What name was given to this "system" of helping slaves escape?

95

Bonus Builder #19

The United States has many famous natural landmarks. Name the state in which each of the following landmarks is located:

a. Grand Canyon   d. Everglades
b. Black Hills   e. Cape Cod
c. Pike's Peak

Can you name a well-known natural landmark in *your* state?

United States Geography

United States Geography

I am on an island in New York Harbor. I am one of the largest statues ever built. The people of France gave me to the United States in 1884. What am I?

United States Geography

Which of the following states have large Hispanic populations? Why?

a. California e. Texas
b. Vermont f. Arizona
c. Mississippi g. Florida
d. New Mexico h. Kentucky

United States Geography

_____ is the southernmost state of the United States.

United States Geography

Which states were the beginning and ending points of the Trail of Tears from 1830 to 1840?

United States Geography

Where did the Oregon Trail begin? Through which states did it pass? Where did the trail end?

Bonus Builder #20

Native Americans built homes with materials they found in their environments. Match each type of Native American home with its location:

a. plank house   Southwest
b. pueblo   Plains
c. tepee   Pacific Coast
d. longhouse   Eastern Woodlands

United States Geography

United States Geography

Which state has the nickname "Mother of Presidents"? Can you list at least five of the presidents born in this state?

101

United States Geography

Which state is divided into two parts: the Upper Peninsula and the Lower Peninsula?

102

United States Geography

Which river forms the boundary between California and Arizona?

103

United States Geography

Which states have *panhandles*? Name at least three.

104

United States Geography

I am part of the United States. However, I am not a state. I am the part of the United States where Christopher Columbus landed. What am I?

105

Bonus Builder #21

The Mississippi River is the second longest river in the United States.
a. In which state is the *source* of the Mississippi River?
b. In which state is the *mouth* of the Mississippi River? What city is at the mouth?

United States Geography

United States Geography

What do all of the following have in common?
President Jimmy Carter
peaches
1996 summer Olympic® Games
Braves
Falcons
Coca-Cola® soda

106

United States Geography

Vermont, New Hampshire, and Rhode Island are all part of what U.S. region? What other three states are included in this region?

107

United States Geography

Which state is farther south: Illinois or Maryland?

108

United States Geography

Which state is made up entirely of islands?

109

United States Geography

In which state is most of Yellowstone National Park located?

110

Bonus Builder #22

The Ozark Plateau is a land region that covers parts of which four states?

United States Geography

United States Geography

I am the second smallest state. Maryland is my neighbor to the west. My state bird is the blue hen chicken. My capital is Dover. What am I?

United States Geography

In which of the following states would you likely find deserts?
  a. South Carolina
  b. Maine
  c. Arizona
  d. Michigan
  e. California
  f. Utah

United States Geography

Which is larger: Great Salt Lake or Lake Erie?

United States Geography

What is the largest city in Missouri? What is the second largest city in Kansas?

United States Geography

Which pair of cities is farther apart: New Orleans and Minneapolis or St. Louis and San Francisco? How did you find out?

List three important U.S. rivers that flow south.

United States Geography

United States Geography

Which region of the United States is known for producing great amounts of wheat?

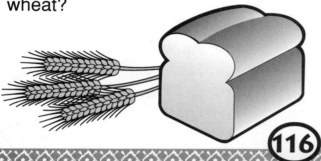

116

United States Geography

Which two states include land that was bought from Mexico as part of the Gadsden Purchase in 1853–54?

117

United States Geography

Which of the following are the top three cotton-growing states?
a. Mississippi
b. Indiana
c. Texas
d. Idaho
e. California
f. Wisconsin

118

United States Geography

In which state is Carlsbad Caverns? What are *caverns*?

119

United States Geography

I am home to the Mile High City. A sports team nicknamed the Broncos plays here. I am west of the Mississippi River. Pikes Peak is located here. What am I?

120

Bonus Builder #24

The Mississippi River is often used to divide the United States into east and west. Label each of the following cities east or west of the Mississippi River:

a. Seattle   d. Houston
b. Memphis   e. Cleveland
c. Denver   f. Phoenix

United States Geography

United States Geography

a. What is the highest mountain peak in the United States? In which state is it located?
b. What is the highest mountain peak east of the Mississippi River? In which state is it located?

121

United States Geography

The *corn belt* is a region that produces large amounts of corn. List the top five corn-producing states.

122

United States Geography

Flying from Wisconsin to Florida in a straight line, what is the fewest number of states over which you would travel (not including Wisconsin and Florida)? List them.

123

United States Geography

Which "state" had its own constitution and governor from 1784 to 1788 but was never admitted to the Union? (Hint: It is now a part of Tennessee.)

124

United States Geography

True or false: The United States is entirely in the west longitudes.

125

Bonus Builder #25

The 40th parallel (40°N latitude) almost divides the United States in half. Which of the following cities are south of the 40th parallel?

   a. Salt Lake City   e. Cincinnati
   b. Denver   f. Pittsburgh
   c. Kansas City, Missouri g. Baltimore
   d. St. Louis   h. San Francisco

United States Geography

United States Geography

List the six time zones of the United States.

126

United States Geography

When it is 9:00 A.M. in San Francisco, what time is it in Honolulu?

127

United States Geography

If it is 5:00 A.M. in Chicago, what time is it in Denver?

128

United States Geography

Is Denver closer to San Francisco or to Chicago?

129

United States Geography

Is Chicago closer to the Atlantic Ocean or to the Pacific Ocean?

130

Bonus Builder #26

Suppose a ship sails from Veracruz, Mexico, to New York City. List two bodies of water over which the ship would sail.

United States Geography

United States Geography

Which state was the first to join the Union? Which state was the last?

131

United States Geography

List the eight states that border Tennessee.

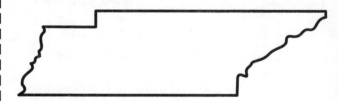

132

United States Geography

Name two states whose names are included in the names of their capital cities.

133

United States Geography

I. Which of the following cities are state capitals?
   a. Sacramento
   b. Tucson
   c. Denver
   d. St. Louis
   e. New Orleans
   f. Chicago
   g. Albany
   h. Boston

II. For the cities above that are not state capitals, identify in what states they are located and what those states' capitals are.

134

United States Geography

I am about one-fifth the size of the rest of the United States. I have the second smallest population. I am home to the highest mountain in North America. What am I?

135

Bonus Builder #27

Every state has a nickname. Match the following states with their nicknames:

   a. Florida   Golden State
   b. New York   Sunshine State
   c. Texas   Empire State
   d. California   Grand Canyon State
   e. Arizona   Lone Star State

United States Geography

WORLD GEOGRAPHY

To travel from London, England, to Paris, France, which of the following forms of transportation could you use?

- a. hovercraft
- b. automobile
- c. train
- d. airplane
- e. ferryboat

(136)

WORLD GEOGRAPHY

Which foreign country, other than Canada and Mexico, is closest to the United States?

(137)

WORLD GEOGRAPHY

What term is used for the region that includes Australia, New Zealand, and more than 25,000 other islands in the Pacific Ocean?

(138)

WORLD GEOGRAPHY

Name the three largest lakes in Africa. Which lake is the chief source of the Nile River?

(139)

WORLD GEOGRAPHY

What money term refers to a U.S. dollar used in the money markets of Europe?

(140)

Bonus Builder #28

Complete the following analogies:

a. Congress is to Washington, DC, as _____ is to London.
b. Washington, DC, is to _____ as London is to England.
c. The United States is to North America as England is to _____.

WORLD GEOGRAPHY

WORLD GEOGRAPHY

Jerusalem is a holy city for three major religions. Name them.

(141)

WORLD GEOGRAPHY

I am an island nation about 90 miles south of Florida. My capital and largest city is Havana. My official language is Spanish. My people call me the "Pearl of the Antilles." What am I?

(142)

WORLD GEOGRAPHY

I am a cereal grain that grows best in shallow water. I am one of the world's most important food crops. China and India are leading producers of me. What am I?

(143)

WORLD GEOGRAPHY

I am an Arab kingdom on the East Bank of the River Jordan. About 95 percent of my people are Muslims. I have deserts, mountains, deep valleys, rolling plains, and a warm, pleasant climate. What am I?

(144)

WORLD GEOGRAPHY

On which continent do Indians, Pakistanis, Chinese, and Mongolians live?

(145)

Bonus Builder #29

Complete the following analogies:
a. Lake Titicaca is to South America as Lake Victoria is to _____.
b. The Amazon River is to _____ as the Nile River is to Africa.
c. The Andes Mountains are to South America as the Alps are to _____.
d. The Gobi is to Asia as _____ is to Africa.

WORLD GEOGRAPHY

WORLD GEOGRAPHY

Match each of the following countries with its major religion:

a. Jordan
b. Japan
c. India
d. Pakistan

Buddhism
Hinduism
Islam

(146)

WORLD GEOGRAPHY

Unscramble each of the following countries. Which one is *not* on the continent of Africa?

a. DAINI
b. IAML
c. DHCA
d. BILYA
e. YEKNA
f. PGEYT

(147)

WORLD GEOGRAPHY

Many geographic names include colors. Name three seas that include colors. In which part of the world is each one located?

(148)

WORLD GEOGRAPHY

I am made up of four large islands and thousands of smaller ones. My population is around 126,000,000 people. Baseball and sumo wrestling are popular sports. My people call me *Nippon,* which means "source of the sun." What am I?

(149)

WORLD GEOGRAPHY

Complete the following analogies:

a. Mount Vesuvius is to Europe as Mount Fuji is to _____.
b. The Rhine River is to Europe as the Tigris River is to _____.
c. The Alps are to _____ as the Himalayas are to Asia.
d. Sicily is to Europe as Sri Lanka is to _____.

(150)

BONUS BUILDER #30

Match each language with its word(s) for *hello*.

a. French   ciao
b. Spanish   hola
c. Italian   jambo
d. German   bonjour
e. Swahili   guten Tag

WORLD GEOGRAPHY

WORLD GEOGRAPHY

I am the second largest island in the Caribbean Sea. Christopher Columbus arrived on my shores in 1492. I am made up of two countries: the Republic of Haiti and the Dominican Republic. My chief crops are coffee, cacao, and sugar cane. What am I?

(151)

WORLD GEOGRAPHY

I am the third largest island in the Caribbean Sea. My citizens speak English. I am one of the world's leading producers of bauxite. My beaches are beautiful! Kingston is my capital and largest city. What am I?

(152)

WORLD GEOGRAPHY

All of the following belong in the same category except one:

a. Kalahari
b. Mojave
c. Great Victoria
d. Painted
e. Pyrenees
f. Gobi
g. Atacama

What is the category? Which one doesn't belong with the others?

(153)

WORLD GEOGRAPHY

I am the most populous country in Africa. My neighbors include Cameroon, Niger, and Benin. I am a large producer of petroleum. English is my official language, but it is not the most commonly used language. What am I?

(154)

WORLD GEOGRAPHY

Which European country is first in alphabetical order? Which one is last? What is the approximate distance between these two countries?

(155)

Bonus Builder #31

Match the European monarch with his or her country.

a. King Harald V
b. King Carl XVI Gustaf
c. Queen Elizabeth II
d. Queen Beatrix
e. Queen Margrethe II

Great Britain and Northern Ireland
Denmark
Norway
Netherlands
Sweden

WORLD GEOGRAPHY

WORLD GEOGRAPHY

I am the only country that is also a continent. I am home to many unusual animals, including the dingo, wallaby, bandicoot, emu, and wombat. I am sometimes referred to as "the land down under." What am I?

156

WORLD GEOGRAPHY

What is Tasmania? Where is Tasmania?

157

WORLD GEOGRAPHY

Which of the following pairs of countries share a common boundary?
a. Algeria and Libya
b. Somalia and Nigeria
c. Mozambique and Gabon
d. Angola and Zambia
e. Egypt and Sierra Leone

158

WORLD GEOGRAPHY

Which of the following cities are *not* in Europe?
a. Madrid
b. Vienna
c. Casablanca
d. Warsaw
e. Amman

159

WORLD GEOGRAPHY

To fly from Bulgaria to the country of Georgia, over which body of water would you cross?

160

Bonus Builder #32

What is the setting for each of the following well-known children's novels?

a. *Owls in the Family*
b. *The Trumpeter of Krakow*
c. *Heidi*
d. *Hans Brinker, or the Silver Skates*
e. *The Secret Garden*

What is *your* favorite novel? What is its setting?

WORLD GEOGRAPHY

WORLD GEOGRAPHY

I am the holiest city in the Muslim religion. I am in western Saudi Arabia. More than a million pilgrims visit here each year. What am I?

161

WORLD GEOGRAPHY

The PLO is a political group that represents the Arab people of Palestine. What does PLO stand for? Who is the leader of the PLO?

162

WORLD GEOGRAPHY

List the following countries in order, greatest to least, by population:

a. China
b. United States
c. Brazil
d. India
e. Indonesia

163

WORLD GEOGRAPHY

Which major river in Africa flows north? Into what body of water does it flow?

164

WORLD GEOGRAPHY

Name an African country that has shorelines on both the Atlantic Ocean and the Mediterranean Sea.

165

Bonus Builder #33

Match each of the following countries with its monetary unit:
a. United Kingdom   dollar
b. Mexico   franc
c. Canada   peso
d. France   pound
e. Israel   mark
f. Germany   shekel

WORLD GEOGRAPHY

World Geography

I was known as Siam until 1939. I am in southeast Asia. Bangkok is my capital and largest city. What am I?

(166)

World Geography

Lapland is located in Europe above the Arctic Circle. Lapland is not a separate country, but is home to a people called the Lapps. Name the four countries that include the region of Lapland within their borders.

(167)

World Geography

On December 31, 1992, Czechoslovakia ceased to exist and became two independent countries. What are they?

(168)

World Geography

If you flew from Montevideo, Uruguay, to Lagos, Nigeria, over which body of water would you fly?

(169)

World Geography

Which river forms part of the boundary between Bulgaria and Romania?

(170)

Bonus Builder #34

List the largest urban area on each continent (except Antarctica). Also list the country in which each urban area is found.

World Geography

WORLD GEOGRAPHY

Name your location if you are at 0° latitude and 0° longitude.

171

WORLD GEOGRAPHY

Name the large, important port city on the western coast of Canada.

172

WORLD GEOGRAPHY

The Maoris are native people of which country?

173

WORLD GEOGRAPHY

Vietnam was a divided nation from 1954 to 1975.

a. What was the capital of North Vietnam?
b. What was the capital of South Vietnam?
c. After North and South Vietnam were unified, what became the capital city?
d. The name of the city that was the capital of South Vietnam was changed. What is its new name?

174

WORLD GEOGRAPHY

There are many types of grasslands throughout the world. On which continent would you find the *Pampas* and the *Llanos*?

175

Bonus Builder #35

Where is each of the following landmarks located?
a. Great Wall
b. Taj Mahal
c. Great Pyramid
d. Eiffel Tower
e. Western (Wailing) Wall
f. Kremlin

WORLD GEOGRAPHY

WORLD GEOGRAPHY

Mountain climbing is a popular sport all over the world. On which continent are you if you're climbing in each of the following mountain ranges?
a. Alps
b. Atlas
c. Himalayas
d. Andes
e. Rocky

(176)

WORLD GEOGRAPHY

List five countries that have scientific stations on Antarctica.

(177)

WORLD GEOGRAPHY

Traveling from Anchorage, Alaska, to Seattle, Washington, through which Canadian province would you pass?

(178)

WORLD GEOGRAPHY

The Río de la Plata is a funnel-shaped bay on the southeastern coast of South America. It's located between Argentina and Uruguay. Which two rivers make up the Río de la Plata?

(179)

WORLD GEOGRAPHY

Which body of water separates Sudan from Saudi Arabia?

(180)

Bonus Builder #36

Every country has a special flag. Label each flag below with its corresponding country.

a. b. c. d. e.

Ghana   Trinidad and Tobago
Israel   Canada
Brazil

WORLD GEOGRAPHY

WORLD GEOGRAPHY

What large saltwater lake in Uzbekistan and Kazakhstan has gotten smaller and smaller due to irrigation?

(181)

WORLD GEOGRAPHY

Name the four countries that make up the United Kingdom.

(182)

WORLD GEOGRAPHY

What part of Egypt lies east of the Suez Canal and the Gulf of Suez, and borders western Israel?

(183)

WORLD GEOGRAPHY

What city is scheduled to host the 2004 summer Olympic® Games?

(184)

WORLD GEOGRAPHY

Sailing from Dar es Salaam, Tanzania, to Bombay, India, over what body of water would you travel?

(185)

Bonus Builder #37

On which continent are all six of the cities in the pairs below?
a. Caracas—Lima   c. Quito—Santiago
b. Brasilia—Montevideo

Match each pair to the approximate distance between the two cities.

  2,400 miles   1,500 miles   1,750 miles

WORLD GEOGRAPHY

WORLD GEOGRAPHY

What are the seasonal winds that blow over the northern part of the Indian Ocean?

(186)

WORLD GEOGRAPHY

Most of the world's volcanoes encircle which ocean? What is this circle of volcanoes often called?

(187)

WORLD GEOGRAPHY

Which Southeast Asian nation is made up of nearly 14,000 islands?

(188)

WORLD GEOGRAPHY

Which two Asian countries have over a third of the world's population?

(189)

WORLD GEOGRAPHY

Which South American country is almost the same size (in area) as the 48 continental United States? Which of these two countries has the larger population?

(190)

Bonus Builder #38

Over time, names of places may change. Match each old name (a–d) with its new name.

a. British Honduras   Myanmar
b. Burma   Belize
c. Rhodesia   Zimbabwe
d. British East Africa   Kenya

WORLD GEOGRAPHY

WORLD GEOGRAPHY

If you flew from Sydney, Australia, to London, England, which route would be shorter: flying west or east from Sydney?

191

WORLD GEOGRAPHY

Which two Canadian provinces have the largest populations?

192

WORLD GEOGRAPHY

List the seven countries of Central America.

193

WORLD GEOGRAPHY

Which South American country has two national capitals? What are they?

194

WORLD GEOGRAPHY

Oceania is the name given to the many thousands of islands scattered across the Pacific Ocean. Geographers have divided these islands into three main groups. One group is called Melanesia. What are the other two?

195

Bonus Builder #39

Match each continent with its longest river.

a. Asia   Yangtze
b. Africa   Murray
c. North America   Volga
d. South America   Missouri
e. Europe   Amazon
f. Australia   Nile

WORLD GEOGRAPHY

WORLD GEOGRAPHY

Nelson Mandela, Cecil John Rhodes, Desmond Tutu, and Frederik de Klerk are important people in the history of which African nation?

196

WORLD GEOGRAPHY

Name Canada's three territories.

197

WORLD GEOGRAPHY

Which country is the northernmost country in South America?

198

WORLD GEOGRAPHY

The following animals are all native to which continent?

pronghorn
coyote
prairie dog
bison

199

WORLD GEOGRAPHY

The following people are important in the history of which North American country?

Montezuma II
Benito Juárez
Maximilian
Diego Rivera

200

Bonus Builder #40

Identify each of the following continents. Then arrange them in land-size order from largest to smallest.

 a.   b.    c.

WORLD GEOGRAPHY

Answer Keys

Page 3
1. Sri Lanka; Ceylon
2. Germany
3. Virginia
4. a. equator c. prime meridian
   b. international date line d. tropic of Capricorn
5. deserts

Bonus Builder #1:
   a. Darwin, Australia
   d. Honolulu, Hawaii

Page 4
6. a. Mount Everest
   b. shores of the Dead Sea
   c. Asia
7. Possible answers: Red River, Missouri River, Ohio River, Arkansas River, Illinois River, Minnesota River, Des Moines River, Rock River, Wisconsin River, Iowa River
8. Carson City, Nevada (Use lines of longitude.)
9. a. Egypt
   b. Panama
   c. New York
10. 1933; Tennessee, North Carolina, Virginia, Georgia, Alabama, Mississippi, and Kentucky

Bonus Builder #2:
   a. north   d. north
   b. south   e. north
   c. north   f. south

Page 5
11. b. South Carolina
    c. Alaska
    e. Florida
    f. Maine
    i. California
12. equator; Arctic Circle
    a. equator
    b. Arctic Circle
13. a. Asia
    b. Asia
    c. Antarctica
14. Australia, Antarctica
15. Answers will vary. A map and a globe are alike in that they both show the earth. They both show similar information through lines, colors, and other symbols. A map and a globe are different because a map is flat and a globe is in the shape of a ball. You cannot see the entire earth at once when looking at a globe, but you can with a map. Because a globe is round like the earth, it represents all parts of the earth's surface true to scale.

Bonus Builder #3: a. Mediterranean Sea
   b. Encounter Bay
   c. Atlantic Ocean
   d. Bay of Bengal
   e. English Channel

Page 6
16. British Columbia, Alberta, Saskatchewan, Manitoba, Ontario, Quebec, New Brunswick
17. a. France
    b. Poland
    c. Germany
    d. Spain
    In order from largest to smallest: France, Spain, Germany, Poland
18. a. west
    b. east
    c. south
    compass rose
19. Texas, Louisiana, Mississippi, Alabama, Florida
20. a. United States/Canada
    b. Venezuela
    c. southern Africa

Bonus Builder #4: Answers will vary. Suggested answers include
   a. Iceland, Ireland, United Kingdom
   b. Japan, Philippines, Sri Lanka, Taiwan
   c. Madagascar

Page 7
21. a. 80 miles
    b. 200 miles
    c. 20 miles
    d. 140 miles
22. A national capital is often indicated by a star within a circle.
    a. Barcelona, c. Marseille, and d. Rotterdam
23. Denmark, Norway, and Sweden
24. b. Paraguay, c. Angola, and f. Japan
25. tropical rain forest

Bonus Builder #5: a. (major) airport
   b. interstate highway
   c. town or city
   d. waterway

Page 8
26. North America, Europe
27. Morocco, Algeria, Tunisia, Libya, and Egypt
28. Asia
29. a. southern tip of South America
    b. between Morocco and Spain
    c. between Alaska and Russia
    d. between England and France
30. the Gobi

Bonus Builder #6: a. physical map
   b. political map
   c. thematic map
   d. thematic map
   e. thematic map

Answer Keys

Page 9
31. Rocky Mountains (Rocky Mountain Chain); New Mexico, Colorado, Utah, Wyoming, Idaho, Montana, Washington, and Alaska
32. Answers may include Snake River, Salmon River, Columbia River, Humboldt River, Colorado River, Gila River, and Green River.
33. Answers will vary. Characteristics of rural communities include less population, fewer services, fewer buildings, less development, more natural areas, and farms. Characteristics of urban communities are lots of development; many services; lots of traffic, noise, and air pollution; industries; and higher population.
34. Similarities of polar regions and deserts: little rainfall, temperature extremes, low population of humans and animals, sparse vegetation, animals and plants adapted to their environment
35. b. strait and e. bight

Bonus Builder #7:
- a. Southern Hemisphere
- b. Northern Hemisphere
- c. Northern Hemisphere
- d. Southern Hemisphere
- e. Northern Hemisphere
- f. Northern Hemisphere

Page 10
36. a. Thames River
    b. Hudson River
    c. Seine River
    d. Mississippi River
    e. Nile River
37. c. Reykjavik
38. the Mediterranean Sea; Africa, Europe, and Asia
39. Chile
40. Russia

Bonus Builder #8: Atlantic Ocean, Strait of Gibraltar, Mediterranean Sea, Suez Canal, Red Sea, Arabian Sea

Page 11
41. United Kingdom
42. Gulf Stream
43. Ural Mountains
44. Australia
45. Andes Mountains

Bonus Builder #9: Bosnia-Herzegovina, Croatia, Macedonia, and Slovenia

Page 12
46. a. tundra
    b. mesa
    d. plateau
    f. island
47. b. Timor
    f. Madagascar
    g. Java
48. South Africa
49. the Sahara
50. north to south; scale of miles

Bonus Builder #10: a. Mozambique Channel
  b. Bering Strait
  c. Bass Strait
  d. English Channel and Strait of Dover

Page 13
51. Answers will vary. Suggestions include title of the map, author of the map, date, compass rose, map key or legend, grid, scale of miles, and symbols.
52. Georgia
53. Greenland
54. Denmark; North America
55. St. Lawrence Seaway

Bonus Builder #11: Albania, Bosnia-Herzegovina, Bulgaria, Macedonia, the mainland of Greece, the European part of Turkey, and parts of Croatia, Slovenia, and Yugoslavia

Page 14
56. along the coastline, especially in the southeast
57. Korea
58. Los Angeles, San Diego, and San Jose
59. east coast of Queensland
60. Newfoundland

Bonus Builder #12: a. Asian Russia
  b. Brazil
  c. European Russia
  d. Sudan
  e. Canada
Australia is the only country that is also a continent. Antarctica has no countries.

Page 15
61. Answers will vary depending on the current season. The seasons in Jerusalem and Tokyo will be the same since they are both in the Northern Hemisphere. The seasons in Montevideo, Buenos Aires, and Sydney will be opposite since they are in the Southern Hemisphere.
62. Answers will vary. Similar qualities: they are environments for plants and animals, names of landforms, can be found throughout the world, can be very hot or very cold, can be in the same places. Differences: mountains have higher altitude; deserts have less rainfall; plants, animals, and people adapt to living in both places differently.
63. Lagos, Nigeria, is warmer because it is closer to the equator.
64. Iceland
65. Africa has the most independent countries (53). The three largest are Sudan, Algeria, and Zaire.

Bonus Builder #13: Estonia, Lithuania, Belarus, Ukraine, Moldova, Kazakhstan, Uzbekistan, Turkmenistan, Georgia, Armenia, Azerbaijan, Kyrgyzstan, Latvia, and Tajikistan

Answer Keys

Page 16
66. Delaware, Maryland, Virginia (which also included the area that became the state of West Virginia in 1863), Kentucky, North Carolina, South Carolina, Georgia, Florida, Alabama, Mississippi, Tennessee, Louisiana, Arkansas, Missouri, and Texas
67. Texas
68. Colorado, Wyoming
69. Benjamin Banneker
70. a. St. Louis, Missouri
    b. Boston, Massachusetts
    c. New York City, New York
    d. Philadelphia, Pennsylvania
    e. Annapolis, Maryland

Bonus Builder #14: a. Netherlands
  b. Sweden
  c. England (and Sweden)
  d. Spain

Page 17
71. a. Italy   c. Mexico
    b. France   d. Japan
72. a. In 1609, Henry Hudson sailed the *Half Moon* up New York's Hudson River. He was searching for the Northwest Passage.
    b. The *Exxon Valdez* tanker ran aground in Alaska's Prince William Sound in March 1989, spilling 11 million gallons of oil.
    c. The British passenger ship *Lusitania* was sunk by a German submarine in the Atlantic Ocean (off the coast of Ireland) on May 7, 1915. A total of 1,198 people died, including 128 Americans.
73. South Dakota
74. a. Florida   d. Texas
    b. New York   e. Nevada
    c. Wisconsin
75. Great Salt Lake

Bonus Builder #15: California (Los Angeles and San Diego) and Texas (Houston, Dallas, and San Antonio)

Page 18
76. a. Tennessee   d. California
    b. South Carolina   e. Arizona
    c. Maryland   f. New York
77. d. Ozark Mountains
78. Alaska, Hawaii
79. a. Austin, Texas   d. Albany, New York
    b. Boston, Massachusetts   e. Salem, Oregon
    c. Boise, Idaho   f. Atlanta, Georgia
80. Answers may vary. Suggestions include Jefferson City, Missouri; Carson City, Nevada; Jackson, Mississippi; Austin, Texas; Madison, Wisconsin; Columbus, Ohio; Lincoln, Nebraska; and Bismarck, North Dakota

Bonus Builder #16: Ohio, Indiana, Illinois, Michigan, Wisconsin, and part of Minnesota. Ohio was the first of these states to join the Union (1803).

Page 19
81. Lake Superior, Lake Huron, Lake Michigan
82. Answers may vary. Suggestions include Ohio, Tennessee, Delaware, Mississippi, Arkansas, Missouri, Illinois, Kansas, and Colorado.
83. Meriwether Lewis and William Clark
84. Washington, DC
85. Trade routes consisted of three paths. The three paths formed a triangle. Example: Rum made in New England was taken to Africa, where it was traded for African slaves. Slaves were taken to the West Indies and traded for molasses. Some slaves stayed on the ships and were taken to the colonies. Molasses from the West Indies was taken back to New England, where it was used to make more rum.

Bonus Builder #17: Minnesota, Iowa, Missouri, Arkansas, and Louisiana

Page 20
86. California, New York, Texas, Florida, and Pennsylvania
87. Alabama, Wyoming; approximately 1,150 miles
88. c. Memphis
    b. Tennessee
    d. United States
    a. North America
89. Navajo
90. a. Santa Fe
    e. Boise
    f. Atlanta

Bonus Builder #18: a. Massachusetts
  b. Virginia
  c. Wisconsin

Page 21
91. a. White House
    b. Arlington National Cemetery
    c. Washington Monument
    d. United States Capitol
92. triangle; Anchorage to Miami
93. a. San Diego, California
    b. Portland, Oregon
    c. Houston, Texas
94. a. chickadee: Maine
    b. nene: Hawaii
    c. brown pelican: Louisiana
    d. cactus wren: Arizona
    e. mountain bluebird: Idaho
95. Underground Railroad

Bonus Builder #19: a. Arizona
  b. South Dakota
  c. Colorado
  d. Florida
  e. Massachusetts

Answer Keys

Page 22
96. Statue of Liberty
97. a. California
    d. New Mexico
    e. Texas
    f. Arizona
    g. Florida
    Many people from Mexico have moved to states along the border. Florida has a large Cuban American population.
98. Hawaii
99. Georgia; Oklahoma
100. Missouri (beginning), Kansas, Nebraska, Wyoming, Idaho, Washington, and Oregon (ending)

Bonus Builder #20: a. Pacific Coast
  b. Southwest
  c. Plains
  d. Eastern Woodlands

Page 23
101. Virginia; George Washington, Thomas Jefferson, James Madison, James Monroe, William Henry Harrison, John Tyler, Zachary Taylor, and Woodrow Wilson
102. Michigan
103. Colorado River
104. Answers may vary. Suggestions include Idaho, Texas, Oklahoma, West Virginia, and Florida.
105. Puerto Rico

Bonus Builder #21: a. Minnesota
  b. Louisiana; New Orleans

Page 24
106. Georgia
107. New England states; Maine, Massachusetts, and Connecticut
108. Illinois
109. Hawaii
110. Wyoming (The park also extends into Idaho and Montana.)

Bonus Builder #22: Arkansas, Illinois, Missouri, and Oklahoma

Page 25
111. Delaware
112. c. Arizona, e. California, and f. Utah
113. Lake Erie
114. Kansas City; Kansas City
115. St. Louis and San Francisco; Use a scale of miles.

Bonus Builder #23: Answers may vary. Suggestions include Missouri River, Mississippi River, Hudson River, and Colorado River.

Page 26
116. the Great Plains
117. Arizona and New Mexico
118. a. Mississippi, c. Texas, and e. California
119. New Mexico; Caverns are large caves.
120. Colorado

Bonus Builder #24: a. west
  b. east
  c. west
  d. west
  e. east
  f. west

Page 27
121. a. Mount McKinley in Alaska, b. Mount Mitchell in North Carolina
122. Iowa, Illinois, Minnesota, Nebraska, and Indiana
123. Four; Illinois, Kentucky, Tennessee, and Alabama
124. Franklin
125. False; Some of Alaska's Aleutian Islands are in the east longitudes.

Bonus Builder #25: b. Denver
  c. Kansas City, Missouri
  d. St. Louis
  e. Cincinnati
  g. Baltimore
  h. San Francisco

Page 28
126. From east to west: Eastern Standard Time, Central Standard Time, Mountain Standard Time, Pacific Standard Time, Alaska Standard Time, and Hawaii-Aleutian Standard Time
127. 7:00 A.M.
128. 4:00 A.M.
129. Chicago
130. Atlantic Ocean

Bonus Builder #26: Gulf of Mexico, Atlantic Ocean

Page 29
131. Delaware; Hawaii
132. Kentucky, Missouri, Arkansas, Mississippi, Alabama, Georgia, North Carolina, and Virginia
133. Indianapolis, Indiana, and Oklahoma City, Oklahoma
134. I. a. Sacramento (California)
       c. Denver (Colorado)
       g. Albany (New York)
       h. Boston (Massachusetts)
    II. b. Arizona: Phoenix
        d. Missouri: Jefferson City
        e. Louisiana: Baton Rouge
        f. Illinois: Springfield
135. Alaska

Bonus Builder #27: a. Sunshine State, b. Empire State, c. Lone Star State, d. Golden State, e. Grand Canyon State

Answer Keys

Page 30
136. All of them. (The tunnel between England and France accommodates shuttle trains that carry passengers and their automobiles.)
137. Russia (only 51 miles from Alaska's most western point)
138. Oceania
139. Lake Victoria, Lake Tanganyika, and Lake Nyasa; Lake Victoria is the chief source of the Nile River.
140. Eurodollar

Bonus Builder #28: a. Parliament
 b. the United States
 c. Europe

Page 31
141. Judaism, Christianity, and Islam
142. Cuba
143. rice
144. Jordan
145. Asia

Bonus Builder #29: a. Africa
 b. South America
 c. Europe
 d. the Sahara

Page 32
146. a. Islam, b. Buddhism, c. Hinduism, d. Islam
147. a. India
 b. Mali
 c. Chad
 d. Libya
 e. Kenya
 f. Egypt
 India is not on the continent of Africa. It is on the continent of Asia.
148. Answers may vary. Yellow Sea—between the east coast of China and Korea; Red Sea—between northeastern Africa and the Arabian Peninsula; Black Sea—western Europe; White Sea—northwest Russia; Coral Sea—northeast coast of Australia
149. Japan
150. a. Asia
 b. Asia
 c. Europe
 d. Asia

Bonus Builder #30: a. bonjour
 b. hola
 c. ciao
 d. guten Tag
 e. jambo

Page 33
151. Hispaniola
152. Jamaica
153. They are all deserts except Pyrenees. The Pyrenees is a mountain chain in Europe.
154. Nigeria
155. Albania, Yugoslavia; Albania and Yugoslavia border each other.

Bonus Builder #31: a. Norway
 b. Sweden
 c. Great Britain and Northern Ireland
 d. Netherlands
 e. Denmark

Page 34
156. Australia
157. Tasmania is the island state of Australia. It is located off the southeast coast of the continent.
158. a. Algeria and Libya; d. Angola and Zambia
159. c. Casablanca (Africa); e. Jordan (Asia)
160. Black Sea

Bonus Builder #32: a. Canada
 b. Poland
 c. Switzerland
 d. Netherlands
 e. England

Page 35
161. Mecca
162. Palestine Liberation Organization; Yasir Arafat
163. a. China, d. India, b. United States, e. Indonesia, and c. Brazil
164. Nile River; Mediterranean Sea
165. Morocco

Bonus Builder #33: a. pound
 b. peso
 c. dollar
 d. franc
 e. shekel
 f. mark

Page 36
166. Thailand
167. Norway, Sweden, Finland, and Russia
168. the Czech Republic and Slovakia
169. Atlantic Ocean
170. Danube River

Bonus Builder #34: Tokyo, Japan (Asia); Mexico City, Mexico (North America); São Paulo, Brazil (South America); Lagos, Nigeria (Africa); Paris, France (Europe); and Sydney, Australia

Answer Keys

Page 37
171. Gulf of Guinea (Atlantic Ocean) off the west coast of Africa
172. Vancouver
173. New Zealand
174. a. Hanoi
    b. Saigon
    c. Hanoi
    d. Ho Chi Minh City
175. South America

Bonus Builder #35: a. China
   b. India
   c. Egypt
   d. France
   e. Israel
   f. Russia

Page 38
176. a. Europe
    b. Africa
    c. Asia
    d. South America
    e. North America
177. Answers may include Argentina, Australia, Chile, Germany, Italy, Japan, New Zealand, Russia, the United Kingdom, and the United States.
178. British Columbia
179. Paraná and Uruguay Rivers
180. Red Sea

Bonus Builder #36: a. Brazil, b. Israel, c. Ghana, d. Canada, e. Trinidad and Tobago

Page 39
181. Aral Sea
182. Scotland, England, Wales, and Northern Ireland
183. Sinai Peninsula
184. Athens, Greece
185. Indian Ocean, Arabian Sea

Bonus Builder #37: South America
   a. 1,750 miles
   b. 1,500 miles
   c. 2,400 miles

Page 40
186. monsoons
187. Pacific Ocean; Ring of Fire
188. Indonesia
189. China and India
190. Brazil; United States

Bonus Builder #38: a. Belize
   b. Myanmar
   c. Zimbabwe
   d. Kenya

Page 41
191. west
192. Ontario and Quebec
193. Belize, Costa Rica, El Salvador, Guatemala, Honduras, Nicaragua, and Panama
194. Bolivia; Sucre and La Paz
195. Micronesia and Polynesia

Bonus Builder #39: a. Yangtze
   b. Nile
   c. Missouri
   d. Amazon
   e. Volga
   f. Murray

Page 42
196. South Africa
197. Yukon, Northwest, and Nunavut
198. Colombia
199. North America
200. Mexico

Bonus Builder #40: a. Africa, b. South America, c. North America; largest to smallest: a. Africa, c. North America, b. South America